Copyright © 2024 by Charlie O. Williams

All rights reserved. No part of this publication may be reproduced, distributed, or transmitted in any form or by any means, including photocopying, recording, or other electronic or mechanical methods, without the prior written permission of the publisher, except in the case of brief quotations embodied in critical reviews and certain other noncommercial uses permitted by copyright law.

For permissions requests, please contact:
Charlie O. Williams

Printed in [United States of America]

First Edition: [May. 2024]

TABLE OF CONTENT

Preface:
Introduction:
Chapter 1: The Origins of Life
Chapter 2: Early Evolutionary Theories
Chapter 3: Darwin and the Theory of Natural Selection
Chapter 4: Mendelian Genetics and Evolution
Chapter 5: Modern Synthesis: Combining Genetic Variation
Chapter 6: Molecular Evolution: DNA and Genetic Variation
Chapter 7: Fossil Record: Tracing the History of Life
Chapter 8: Extinction Events and Mass Extinctions
Chapter 9: Phylogenetics: Reconstructing Evolutionary Relationships
Chapter 10: Adaptive Radiation: Diversification of Species
Chapter 11: Coevolution: Interactions Between Species
Chapter 12: Evolutionary Developmental Biology

Chapter 13: Speciation: Origin of New Species
Chapter 14: Human Evolution: From Australopithecus to Homo sapiens
Chapter 15: Evolutionary Medicine: Applying Evolutionary Principles to Health
Chapter 16: Conservation Biology: Evolutionary Approaches to Biodiversity Conservation
Chapter 17: Evolutionary Ecology: Interactions Between Organisms and Their Environment
Conclusion: Exploring the Tapestry of Life
Acknowledgements

Evolutionary Biology: Tracing the History of Life on Earth

**Author
Charlie O. Williams**

Preface:

In the vast expanse of time and space, the story of life on Earth unfolds with unparalleled complexity and beauty. As we delve into the pages of "Evolutionary Biology: Tracing the History of Life on Earth," we embark on a captivating journey through the annals of natural history. This book serves as a tribute to the remarkable journey of discovery undertaken by generations of scientists who have tirelessly pursued the secrets of evolution. From the humble beginnings of life to the dazzling array of species that inhabit our planet today, each chapter in this volume illuminates a different facet of the evolutionary process. Whether you are a seasoned biologist, an aspiring student, or simply a curious explorer of the natural world, we invite you to join us on this exhilarating odyssey through the depths of evolutionary biology.

Introduction:

Evolutionary biology stands as a cornerstone in our understanding of life on Earth, offering profound insights into the intricate tapestry of biological diversity that surrounds us. From the emergence of the first single-celled organisms to the complex web of life we see today, the field of evolutionary biology provides a roadmap for tracing the history of life on our planet. In this comprehensive exploration, we embark on a journey through time, unraveling the mysteries of how species evolve, adapt, and thrive in ever-changing environments. Join us as we delve into the fundamental principles, groundbreaking discoveries, and ongoing debates that shape our understanding of the evolutionary process.

Chapter 1: The Origins of Life

In the vast expanse of time, there exists a profound mystery that has captivated the minds of scientists for centuries: the origins of life. To trace the history of life on Earth, we must embark on a journey back to the very beginning, to a primordial world teeming with potential and possibility.

The Primordial Soup

Our story begins approximately 4.5 billion years ago, in a young and volatile Earth. At this time, the planet was a chaotic crucible of molten rock, bombarded by asteroids and comets. Yet, amidst this turmoil, there existed a glimmer of hope for life to emerge.

One prevailing hypothesis for the origin of life is the concept of the "primordial soup." Proposed by Soviet biologist Alexander Oparin in the 1920s and later refined by Stanley Miller and Harold Urey in their famous experiment in 1952, this theory suggests that the early Earth's atmosphere was rich in gases such as methane, ammonia, hydrogen, and water vapor. These gases, subjected to intense energy sources such as lightning storms and ultraviolet radiation, formed simple organic molecules, including amino acids—the building blocks of proteins.

The RNA World

While the synthesis of organic molecules was a crucial step, the transition from non-living to living entities required the emergence of self-replicating molecules. Enter ribonucleic acid, or RNA.

In the 1960s, biologist Carl Woese proposed the concept of the "RNA world," suggesting that RNA molecules played a central role in the origin of life. Unlike DNA, which stores genetic information, and proteins, which carry out cellular functions, RNA possesses both catalytic and informational properties. This versatility makes RNA a prime candidate for the first self-replicating molecule.

The Origin of Cells

But how did these primitive molecules organize themselves into the first living cells? The answer lies in the formation of protocells—simple, membrane-bound structures that exhibit some characteristics of living cells.

Research conducted by scientists like Jack Szostak and Pier Luigi Luisi has shed light on the possible mechanisms by which protocells could have formed. These researchers have demonstrated that simple lipid membranes can spontaneously assemble in aqueous environments, enclosing RNA molecules and other biomolecules within their boundaries. Over time, these protocells could have evolved the capacity for metabolism, growth, and division, laying the groundwork for the first true cells.

The Search for Life Beyond Earth

As we unravel the mysteries of life's origins on Earth, our quest for understanding extends beyond the confines of our own planet. Scientists are actively searching for signs of life elsewhere in the universe, exploring the potential for life on other planets and moons within our solar system, as well as exoplanets orbiting distant stars.

The discovery of extremophiles—organisms capable of surviving in extreme environments such as deep-sea hydrothermal vents, acidic hot springs, and Antarctic ice—has expanded our understanding of the conditions under which life can thrive. These findings suggest that life may be more resilient and adaptable than previously thought, opening up new possibilities for the existence of extraterrestrial life.

The origins of life remain one of the greatest scientific puzzles of our time, a tantalizing enigma that continues to inspire researchers around the world. As we delve deeper into the complexities of abiogenesis—the transition from non-life to life—we gain not only a greater appreciation for the remarkable journey that has led to the diversity of life on Earth but also a glimpse into the potential for life to exist elsewhere in the cosmos. In the chapters that follow, we will explore how life has evolved and diversified over billions of years, shaping the world we inhabit today. But first, we must pay homage to the humble beginnings of life itself, rooted in the primordial soup of our planet's distant past.

Chapter 2: Early Evolutionary Theories

As humanity gazed upon the wonders of the natural world, questions about the diversity of life and its origins inevitably arose. In this chapter, we explore the early evolutionary theories that laid the foundation for our modern understanding of biology.

The Ancient Greeks: Seeds of Inquiry

The roots of evolutionary thought can be traced back to ancient civilizations, where philosophers and naturalists pondered the origins of life and the diversity of species. Among these early thinkers, the ancient Greeks made significant contributions to our understanding of the natural world.

Aristotle, often regarded as the father of biology, proposed a scala naturae, or "scale of nature," in which living organisms were arranged in a hierarchical order from simplest to most complex. While Aristotle's ideas were influential in shaping Western thought, they did not explicitly address the concept of evolution.

Lamarckism: The Inheritance of Acquired Characteristics

In the early 19th century, French naturalist Jean-Baptiste Lamarck proposed one of the first comprehensive theories of evolution. Lamarck suggested that organisms could evolve over time through the inheritance of acquired characteristics. According to Lamarck, individuals could acquire new traits during their lifetime in response to environmental pressures, and these acquired traits could be passed on to their offspring.

While Lamarck's ideas were groundbreaking in their time, they were later supplanted by Charles Darwin's theory of natural selection. Nevertheless, Lamarck's emphasis on the role of the environment in shaping evolution foreshadowed later developments in ecological and evolutionary biology.

Darwin's Precursors: The Road to Natural Selection

Before Darwin's groundbreaking work on evolution, several scientists laid the groundwork for his theory of natural selection. One such figure was the Scottish geologist James Hutton, who proposed the principle of uniformitarianism—the idea that the Earth's geological features could be explained by processes that are still operating today.

Another influential figure was the English economist Thomas Malthus, whose essay on population dynamics inspired Darwin to formulate his theory of natural selection. Malthus argued that populations have the potential to grow exponentially, but resources are limited, leading to competition for survival.

Darwin and the Theory of Natural Selection

In 1859, Charles Darwin published his seminal work, "On the Origin of Species," in which he proposed the theory of natural selection as the mechanism driving evolution. Darwin's theory, influenced by his observations during his voyage on the HMS Beagle and his studies of domestic breeding, revolutionized our understanding of the natural world.

According to Darwin, individuals within a population exhibit variation, and those variations that confer advantages in terms of survival and reproduction are more likely to be passed on to future generations. Over time, this process of natural selection leads to the accumulation of beneficial traits and the divergence of species from a common ancestor.

The early evolutionary theories explored in this chapter represent the beginning of humanity's quest to unravel the mysteries of life's diversity and origins. From the speculative musings of ancient philosophers to the rigorous observations of naturalists and scientists, each contribution has played a role in shaping our understanding of the natural world. In the chapters that follow, we will delve deeper into the mechanisms of evolution, exploring how Darwin's theory of natural selection laid the groundwork for modern evolutionary biology.

Chapter 3: Darwin and the Theory of Natural Selection

In the annals of scientific history, few figures loom as large as Charles Darwin. His theory of natural selection revolutionized our understanding of the natural world and forever altered the course of biology. In this chapter, we delve into the life and work of Charles Darwin, tracing the development of his revolutionary theory of evolution by natural selection.

Early Life and Influences

Charles Robert Darwin was born on February 12, 1809, in Shrewsbury, England, into a family of prominent physicians and intellectuals. From an early age, Darwin showed an interest in the natural world, collecting specimens and conducting experiments in his father's garden.

Darwin's passion for science led him to enroll at the University of Edinburgh to study medicine, but he soon found the lectures dull and the surgical procedures gruesome. He transferred to the University of Cambridge to study theology, where he became friends with the botanist John Stevens Henslow, who would later become a mentor and influence on Darwin's scientific career.

The Voyage of the Beagle

In 1831, Darwin embarked on a fateful journey aboard the HMS Beagle as a naturalist and companion to Captain Robert FitzRoy. The voyage, which lasted nearly five years, took Darwin to remote regions of South America, the Galápagos Islands, Australia, and other parts of the world.

During his travels, Darwin made a series of groundbreaking observations that would shape his thinking about the origin and diversity of species. In the Galápagos Islands, he noted subtle variations in the beaks of finches and the unique characteristics of tortoises and other species on different islands—a discovery that would later form the basis of his theory of natural selection.

Formulating the Theory of Natural Selection

After returning from his voyage on the Beagle, Darwin spent years meticulously collecting and analyzing data to support his ideas about evolution. In 1858, Darwin received a letter from Alfred Russel Wallace, a naturalist working in the Malay Archipelago, outlining a similar theory of evolution by natural selection.

Realizing that Wallace had independently arrived at the same conclusion, Darwin hastened to publish his own work, "On the Origin of Species," in 1859. In this groundbreaking book, Darwin presented his theory of natural selection as the mechanism driving evolution—a theory supported by extensive evidence from the fossil record, comparative anatomy, embryology, and biogeography.

Controversy and Legacy

Darwin's theory of natural selection sparked immediate controversy and debate, challenging long-held religious and philosophical beliefs about the origins of life. Yet, over time, his ideas gained widespread acceptance within the scientific community and revolutionized the field of biology.

Darwin's legacy extends far beyond his theory of natural selection. His work laid the groundwork for modern evolutionary biology and influenced diverse fields such as genetics, ecology, and anthropology. Today, Darwin is revered as one of the greatest scientists in history, and his theory of evolution by natural selection stands as a testament to the power of human curiosity and intellect in unraveling the mysteries of the natural world.

Charles Darwin's theory of natural selection represents a landmark achievement in the history of science, reshaping our understanding of life's origins and diversity. From his humble beginnings in Shrewsbury to his historic voyage aboard the HMS Beagle, Darwin's life was a testament to the power of observation, curiosity, and perseverance. In the chapters that follow, we will explore how Darwin's theory laid the foundation for modern evolutionary biology and continue to shape our understanding of life on Earth.

Chapter 4: Mendelian Genetics and Evolution

In the late 19th century, the field of biology witnessed a convergence of two revolutionary ideas: Mendelian genetics and Darwinian evolution. The synthesis of these concepts laid the foundation for modern evolutionary biology and transformed our understanding of how traits are inherited and how species evolve. In this chapter, we explore the intersection of Mendelian genetics and evolution and the profound implications of this synthesis for our understanding of life on Earth.

Mendel's Peas: The Birth of Genetics

Our story begins with Gregor Mendel, an Augustinian friar and scientist who conducted groundbreaking experiments on pea plants in the mid-19th century. Mendel meticulously cross-pollinated pea plants with different traits, such as flower color and seed shape, and carefully tracked the inheritance patterns of these traits over multiple generations.

Mendel's experiments revealed the existence of discrete units of inheritance, which we now know as genes, and the principles of segregation and independent assortment. His work laid the foundation for modern genetics and provided crucial insights into the mechanisms of heredity.

The Rediscovery of Mendel's Work

Despite the significance of Mendel's discoveries, his work went largely unnoticed by the scientific community until it was independently rediscovered by botanists Carl Correns, Hugo de Vries, and Erich von Tschermak in the early 20th century. These researchers recognized the importance of Mendel's experiments and helped to popularize his ideas within the scientific community.

The rediscovery of Mendel's work provided a crucial missing piece of the puzzle for evolutionary biologists, who were grappling with the mechanisms of inheritance and variation. It paved the way for the synthesis of Mendelian genetics and Darwinian evolution, a synthesis that would revolutionize our understanding of the natural world.

Mendelian Inheritance and Evolution

The principles of Mendelian inheritance have profound implications for evolutionary biology. By elucidating the mechanisms of heredity, Mendel's work provided a theoretical framework for understanding how traits are passed from one generation to the next and how new variations arise within populations.

One of the key insights from Mendelian genetics is the concept of alleles—alternative forms of a gene that can produce different phenotypic traits. Variation in alleles within a population is the raw material for evolution, providing the genetic diversity upon which natural selection can act.

The Modern Synthesis

The synthesis of Mendelian genetics and Darwinian evolution, often referred to as the modern synthesis, occurred in the first half of the 20th century. This synthesis was spearheaded by scientists such as Ronald Fisher, J.B.S. Haldane, and Sewall Wright, who integrated Mendelian genetics with Darwin's theory of natural selection.

The modern synthesis provided a comprehensive framework for understanding how genetic variation arises, how it is maintained within populations, and how it leads to evolutionary change over time. It laid the groundwork for modern evolutionary biology and remains the foundation of our current understanding of how species evolve.

The synthesis of Mendelian genetics and Darwinian evolution represents a pivotal moment in the history of biology, bringing together two disparate fields of study and forging a unified theory of how life evolves. From Mendel's humble experiments with pea plants to the modern synthesis of genetics and evolution, our understanding of the natural world has been profoundly shaped by the insights of these visionary scientists. In the chapters that follow, we will continue to explore the mechanisms of evolution and the remarkable diversity of life on Earth.

Chapter 5: Modern Synthesis: Combining Genetic Variation

In the early 20th century, the fields of genetics and evolutionary biology stood at a crossroads.

While Mendelian genetics had provided crucial insights into the mechanisms of heredity, Darwin's theory of natural selection offered a powerful explanation for the origin of species. In this chapter, we explore how the modern synthesis brought these two fields together, revolutionizing our understanding of how genetic variation drives evolution.

The Need for Integration

As the study of genetics and evolutionary biology advanced, it became increasingly clear that a synthesis of these two disciplines was necessary to fully understand the mechanisms of evolution. While Mendelian genetics provided a framework for understanding the transmission of traits from one generation to the next, Darwinian evolution focused on the processes that shape the distribution of traits within populations over time.

The modern synthesis sought to reconcile these seemingly disparate perspectives, integrating the principles of genetics with the mechanisms of natural selection to create a unified theory of evolution.

The Architects of the Modern Synthesis

The modern synthesis was shaped by the contributions of a diverse group of scientists from around the world. British geneticist Ronald Fisher, for example, applied mathematical models to population genetics, demonstrating how natural selection could act on genetic variation within populations.

J.B.S. Haldane, another British scientist, made significant contributions to the understanding of genetic linkage and the role of mutation in evolution. His work laid the groundwork for the concept of the "genetic load," or the accumulation of deleterious mutations in a population.

American biologist Sewall Wright introduced the concept of genetic drift, highlighting the role of random chance in shaping genetic variation within populations. Wright's research emphasized the importance of population size and structure in determining the trajectory of evolution.

Population Genetics and Evolutionary Dynamics

Central to the modern synthesis is the field of population genetics, which examines how genetic variation changes over time within populations. Population genetics provides a theoretical framework for understanding how evolutionary forces such as natural selection, genetic drift, migration, and mutation shape patterns of genetic variation within and between populations.

One of the key insights from population genetics is the concept of the "gene pool," or the total collection of genes and alleles within a population. Changes in the gene pool over time, driven by evolutionary forces, lead to the adaptation and diversification of species.

Applications of the Modern Synthesis

The modern synthesis has had far-reaching implications for our understanding of biology and has found applications in diverse fields such as medicine, agriculture, and conservation. By elucidating the mechanisms of evolution, the modern synthesis has provided insights into the origins of genetic diseases, the breeding of crops and livestock, and the conservation of endangered species.

Furthermore, advances in molecular biology and genomics have allowed scientists to study genetic variation at the molecular level, providing unprecedented insights into the processes of evolution. Techniques such as DNA sequencing and genome editing have opened up new avenues for research and discovery, further enriching our understanding of the natural world.

The modern synthesis represents a triumph of scientific inquiry, bringing together diverse fields of study to create a unified theory of evolution. By integrating the principles of genetics with the mechanisms of natural selection, the modern synthesis has provided a comprehensive framework for understanding how genetic variation drives the diversity of life on Earth. In the chapters that follow, we will explore how these principles have shaped the evolution of species and the emergence of biological diversity.

Chapter 6: Molecular Evolution: DNA and Genetic Variation

As scientists delved deeper into the mechanisms of evolution, they turned their attention to the molecular level, seeking to understand how genetic variation manifests in the building blocks of life: DNA. In this chapter, we explore the field of molecular evolution, which has revolutionized our understanding of how species evolve and diversify over time.

The Discovery of DNA

The story of molecular evolution begins with the discovery of DNA, the molecule that carries the genetic instructions for all living organisms. In 1953, James Watson and Francis Crick elucidated the double-helix structure of DNA, based on the work of Rosalind Franklin and others. This discovery provided the foundation for our understanding of how genetic information is stored and transmitted from one generation to the next.

Genetic Variation and Mutation

At the heart of molecular evolution lies the concept of genetic variation—the diversity of alleles and genes within a population. Genetic variation arises through several mechanisms, including mutation, recombination, and gene flow.

Mutations, in particular, play a crucial role in generating genetic diversity. These spontaneous changes in the DNA sequence can result from errors during DNA replication, exposure to mutagenic agents, or the insertion of foreign DNA sequences. While most mutations are neutral or deleterious, some can confer advantages in certain environments, driving evolutionary change.

Molecular Clocks and Phylogenetics

One of the key insights from molecular evolution is the concept of molecular clocks, which use the rate of molecular change to estimate the timing of evolutionary events. By comparing DNA sequences from different species, scientists can infer the evolutionary relationships between them and estimate the divergence times of common ancestors.

Phylogenetics, the study of evolutionary relationships, relies heavily on molecular data to reconstruct the branching patterns of the tree of life. Techniques such as DNA sequencing and bioinformatics have revolutionized the field, allowing researchers to analyze vast amounts of genetic information and infer the evolutionary history of organisms with unprecedented accuracy.

Molecular Adaptation and Evolutionary Dynamics

Molecular evolution also provides insights into the adaptive processes that drive evolution at the molecular level. By studying the selective pressures acting on genes and proteins, scientists can identify regions of the genome that have undergone positive selection—indicating adaptation to specific environmental conditions or ecological niches.

Furthermore, molecular evolution allows us to investigate the mechanisms of genetic drift, gene flow, and genetic hitchhiking, which shape the distribution of genetic variation within and between populations. These evolutionary dynamics play a central role in shaping the genetic diversity of species and driving the processes of speciation and adaptation.

Applications of Molecular Evolution

The insights gained from molecular evolution have wide-ranging applications in fields such as medicine, agriculture, and conservation. By understanding the genetic basis of diseases, scientists can develop targeted therapies and diagnostic tools to improve human health. In agriculture, molecular evolution informs breeding programs aimed at enhancing crop yields and resilience to environmental stresses.

Furthermore, molecular evolution provides valuable tools for conservation biologists, allowing them to assess genetic diversity within endangered populations and design strategies for preserving biodiversity. Techniques such as DNA barcoding and population genomics have revolutionized our ability to monitor and manage threatened species and ecosystems.

Molecular evolution has transformed our understanding of how genetic variation drives the processes of evolution and diversification. By elucidating the molecular mechanisms of inheritance, mutation, and adaptation, scientists have gained unprecedented insights into the origins and dynamics of life on Earth. In the chapters that follow, we will continue to explore how these principles shape the evolution of species and the emergence of biological diversity.

Chapter 7: Fossil Record: Tracing the History of Life

In the quest to unravel the mysteries of evolution, few sources of evidence are as rich and revealing as the fossil record. Fossils provide a window into the distant past, allowing scientists to trace the history of life on Earth and reconstruct the evolutionary journeys of organisms long gone. In this chapter, we explore the fossil record and its profound implications for our understanding of evolutionary biology.

The Origins of Paleontology

The study of fossils, known as paleontology, has its roots in ancient civilizations, where fossils were often regarded as curiosities or divine artifacts. However, it was not until the 17th and 18th centuries that paleontology began to emerge as a scientific discipline.

In the 17th century, Danish naturalist Nicolaus Steno made significant contributions to the field by recognizing the organic origins of fossils and developing principles for interpreting the relative ages of rock layers. Steno's work laid the groundwork for the concept of stratigraphy, which forms the basis of modern geological dating methods.

Mary Anning and the Fossil Hunters

In the early 19th century, the pioneering work of fossil hunters such as Mary Anning helped to revolutionize our understanding of prehistoric life. Anning, a self-taught paleontologist from Lyme Regis, England, discovered numerous fossils of marine reptiles, including ichthyosaurs and plesiosaurs, along the Jurassic Coast.

Anning's discoveries provided crucial evidence for the existence of extinct species and challenged prevailing beliefs about the fixity of species. Her work paved the way for future generations of paleontologists to explore the depths of Earth's history.

Evolutionary Patterns in the Fossil Record

The fossil record offers a rich tapestry of evolutionary patterns, revealing the rise and fall of ancient organisms and the transitions between major groups of life. Through careful study and analysis, scientists have identified key evolutionary events, such as the Cambrian explosion, which witnessed the rapid diversification of multicellular life forms over 500 million years ago.

Fossils also provide evidence for evolutionary processes such as speciation, adaptive radiation, and extinction. By examining the distribution of fossils in different geological layers, scientists can infer the tempo and mode of evolutionary change and reconstruct the evolutionary relationships between different groups of organisms.

Transitional Fossils and Evolutionary Transitions

One of the most compelling lines of evidence for evolution comes from transitional fossils—organisms that exhibit characteristics of both ancestral and descendant species. These fossils provide snapshots of evolutionary transitions, offering insights into the gradual changes that have occurred over millions of years.

Transitional fossils have played a crucial role in documenting major evolutionary transitions, such as the transition from fish to tetrapods, reptiles to mammals, and land-dwelling mammals to whales. These fossils provide tangible evidence for the evolutionary processes described by Darwin and support the concept of common descent among living organisms.

Challenges and Controversies

While the fossil record provides invaluable insights into the history of life on Earth, it is not without its challenges and controversies. Gaps in the fossil record, known as stratigraphic or temporal discontinuities, can obscure our understanding of evolutionary transitions and the timing of key events.

Furthermore, interpreting fossils can be a complex and subjective endeavor, requiring careful consideration of factors such as taphonomy, preservation bias, and the fidelity of the fossilization process. Despite these challenges, advances in techniques such as computed tomography (CT) scanning and synchrotron imaging have enhanced our ability to study fossils in unprecedented detail.

The fossil record stands as a testament to the enduring legacy of life on Earth, providing a window into the evolutionary processes that have shaped the diversity of organisms we see today. From the humble beginnings of single-celled organisms to the majestic dinosaurs of the Mesozoic Era, fossils offer a glimpse into the grand sweep of Earth's history. In the chapters that follow, we will continue to explore the fossil record and its implications for our understanding of evolutionary biology.

Chapter 8: Extinction Events and Mass Extinctions

Throughout the history of life on Earth, species have come and gone, but some extinction events have left indelible marks on the planet's biodiversity. In this chapter, we explore the causes and consequences of extinction events, with a particular focus on mass extinctions—the cataclysmic events that have reshaped the course of evolution.

Understanding Extinction

Extinction is a natural process that occurs when species cease to exist either locally or globally. While extinction events can result from a variety of factors, including competition, habitat loss, and environmental change, they are often precipitated by sudden and catastrophic events.

The fossil record provides a rich source of evidence for past extinction events, allowing scientists to study the patterns and causes of extinction over geological time scales. By understanding the mechanisms of extinction, scientists can gain insights into the resilience of species and ecosystems in the face of environmental challenges.

The Big Five: Mass Extinction Events

Mass extinctions represent the most severe and far-reaching extinction events in Earth's history, resulting in the loss of a significant portion of the planet's biodiversity. While extinction is an ongoing process, mass extinctions are characterized by a rapid and widespread loss of species across multiple taxonomic groups.

There have been five major mass extinctions in Earth's history, each associated with distinct geological events and environmental changes. These include the end-Ordovician, Late Devonian, end-Permian, end-Triassic, and end-Cretaceous mass extinctions, which occurred over the past 500 million years.

The End-Permian Extinction: The Great Dying

The end-Permian extinction, which occurred approximately 252 million years ago, stands as the most severe extinction event in Earth's history. Up to 96% of marine species and 70% of terrestrial vertebrate species went extinct during this event, leading to profound changes in the composition of life on Earth.

The causes of the end-Permian extinction are still debated, but hypotheses include massive volcanic eruptions, asteroid impacts, and climate change. Regardless of the exact trigger, the end-Permian extinction fundamentally altered the course of evolution, paving the way for the rise of new species and ecosystems in the aftermath.

The Cretaceous-Paleogene Extinction: The Demise of the Dinosaurs

The end-Cretaceous extinction, which occurred approximately 66 million years ago, is perhaps the most famous extinction event in Earth's history. This event marked the end of the Mesozoic Era and the demise of the dinosaurs, as well as numerous other plant and animal species.

The leading hypothesis for the end-Cretaceous extinction is the impact of a large asteroid or comet, which triggered widespread environmental devastation, including wildfires, tsunamis, and a global "nuclear winter." This catastrophic event paved the way for the rise of mammals and birds as dominant terrestrial vertebrates and ultimately led to the evolution of modern ecosystems.

The Legacy of Mass Extinctions

Mass extinctions have left a lasting imprint on the planet's biodiversity, shaping the course of evolution and influencing the patterns of life we see today. While these events are often associated with devastation and loss, they also create opportunities for new species to evolve and thrive in the aftermath.

Furthermore, mass extinctions serve as cautionary tales about the fragility of life and the interconnectedness of species and ecosystems. By studying past extinction events, scientists can gain insights into the potential impacts of ongoing environmental changes and develop strategies for conserving biodiversity and mitigating the effects of future extinctions.

Extinction events and mass extinctions are integral parts of Earth's evolutionary history, shaping the diversity of life on the planet over millions of years. From the catastrophic events that led to the demise of the dinosaurs to the subtle shifts in biodiversity that occur over geological time scales, extinction events remind us of the dynamic and ever-changing nature of life on Earth. In the chapters that follow, we will continue to explore the processes of evolution and adaptation that drive the persistence and diversification of species in the face of environmental challenges.

Chapter 9: Phylogenetics: Reconstructing Evolutionary Relationships

In the quest to understand the tree of life, scientists turn to phylogenetics—a powerful tool for reconstructing the evolutionary relationships between species. In this chapter, we delve into the principles and methods of phylogenetics, exploring how this field has transformed our understanding of evolutionary biology.

The Foundations of Phylogenetics

The study of phylogenetics has its roots in the work of early naturalists and taxonomists who sought to classify and organize the diversity of life on Earth. Swedish botanist Carl Linnaeus, for example, developed the binomial nomenclature system for naming species, laying the groundwork for modern taxonomy.

The concept of phylogeny—the evolutionary history of a group of organisms—emerged in the 19th century, as scientists began to recognize patterns of similarity and divergence among species. However, it was not until the advent of molecular genetics in the 20th century that phylogenetics truly began to flourish as a field of study.

Building the Tree of Life

At the heart of phylogenetics lies the construction of phylogenetic trees—diagrams that depict the evolutionary relationships between species. Phylogenetic trees are based on the principle of common descent, which posits that all living organisms share a common ancestor and have diversified through a process of descent with modification.

Phylogenetic trees are constructed using a variety of methods, including morphological, molecular, and behavioral data. Molecular phylogenetics, in particular, relies on DNA sequences to infer evolutionary relationships, providing a powerful tool for studying the evolutionary history of organisms.

Methods of Phylogenetic Inference

Phylogenetic inference involves the reconstruction of ancestral relationships based on observed patterns of similarity and divergence among species. One common approach is the cladistic method, which identifies shared derived traits, or synapomorphies, that unite groups of organisms into clades.

Another widely used method is maximum likelihood, which uses statistical models to estimate the probability of different evolutionary scenarios given the observed data. Maximum likelihood allows researchers to account for factors such as mutation rates, genetic drift, and selection pressure when reconstructing phylogenetic trees.

Applications of Phylogenetics

Phylogenetics has a wide range of applications in fields such as evolutionary biology, ecology, biogeography, and conservation. By reconstructing phylogenetic trees, scientists can gain insights into the evolutionary processes that have shaped the diversity of life on Earth and identify key evolutionary innovations and transitions.

Phylogenetic methods also play a crucial role in biodiversity conservation, allowing researchers to prioritize conservation efforts based on evolutionary distinctiveness and evolutionary history. By identifying evolutionary hotspots and areas of endemism, conservation biologists can develop strategies for protecting species and ecosystems with high conservation value.

Challenges and Future Directions

While phylogenetics has revolutionized our understanding of evolutionary relationships, it is not without its challenges and limitations. Factors such as incomplete taxon sampling, gene duplication and loss, and horizontal gene transfer can complicate phylogenetic inference and lead to erroneous conclusions.

Furthermore, the increasing availability of genomic data has led to the development of more complex models of evolution, raising questions about how best to incorporate this information into phylogenetic analyses. Despite these challenges, advances in computational methods and statistical modeling continue to push the boundaries of phylogenetics, allowing scientists to explore the tree of life with unprecedented detail and precision.

Phylogenetics stands as a cornerstone of evolutionary biology, providing a powerful framework for understanding the evolutionary relationships between species. From the earliest pioneers of taxonomy to the modern era of molecular genetics, phylogenetics has transformed our understanding of the tree of life and continues to shape the way we study and interpret the diversity of life on Earth. In the chapters that follow, we will explore how phylogenetic methods have been used to unravel the mysteries of evolution and uncover the hidden connections between species across the tree of life.

Chapter 10: Adaptive Radiation: Diversification of Species

One of the most remarkable phenomena in the history of life on Earth is adaptive radiation—the rapid diversification of species from a common ancestor into a variety of ecological niches. In this chapter, we explore the processes driving adaptive radiation and the remarkable examples of evolutionary innovation that have shaped the diversity of life.

Understanding Adaptive Radiation

Adaptive radiation occurs when a single ancestral species undergoes rapid speciation to exploit a wide range of ecological opportunities. These opportunities may arise due to the colonization of new habitats, the extinction of competitors, or the evolution of key innovations that allow organisms to exploit new resources.

The classic example of adaptive radiation is the finches of the Galápagos Islands, which diversified into a variety of forms with specialized beak shapes to exploit different food sources. Adaptive radiation can occur in a variety of environments, from islands and archipelagos to lakes, forests, and deserts.

Darwin's Finches: A Case Study in Adaptive Radiation

The finches of the Galápagos Islands, famously studied by Charles Darwin during his voyage on the HMS Beagle, provide a textbook example of adaptive radiation in action. These finches, which belong to the genus Geospiza, diversified into a variety of species with different beak shapes and feeding habits.

Darwin observed that the finches' beak shapes were correlated with the types of food available on each island, with larger, stronger beaks adapted for cracking seeds and smaller, more delicate beaks adapted for feeding on insects and vegetation. This diversification allowed the finches to exploit a wide range of ecological niches and thrive in the harsh and variable environments of the Galápagos.

Key Innovations and Ecological Opportunity

Adaptive radiation is often facilitated by the evolution of key innovations—traits or adaptations that allow organisms to exploit new resources or environments. These innovations may include anatomical features such as wings, limbs, or beaks, as well as behavioral traits such as mating displays or foraging strategies.

For example, the evolution of flight in birds and insects opened up new opportunities for colonization and diversification, allowing these organisms to exploit aerial habitats and escape predators. Similarly, the evolution of jaws and teeth in early vertebrates facilitated the colonization of new aquatic and terrestrial environments, leading to the diversification of modern vertebrate groups.

Patterns of Adaptive Radiation

Adaptive radiation can lead to a variety of patterns of species diversification, depending on the ecological opportunities available and the constraints imposed by the environment. In some cases, adaptive radiation may result in the rapid proliferation of species into a wide range of niches, leading to a "burst" of diversification.

In other cases, adaptive radiation may be more gradual, with species diversifying over longer periods of time in response to changing environmental conditions. Regardless of the pattern, adaptive radiation often results in the formation of species flocks—groups of closely related species that share a common ancestor and occupy similar ecological niches.

Convergence and Parallel Evolution

One of the striking features of adaptive radiation is the phenomenon of convergence, where distantly related species evolve similar traits in response to similar ecological pressures. Convergent evolution often results in the independent evolution of similar morphological, physiological, or behavioral adaptations in unrelated lineages.

A classic example of convergence is seen in the marsupial mammals of Australia and the placental mammals of other continents, which have evolved similar body shapes and ecological roles despite their evolutionary separation. Similarly, the cactus-like forms of cacti in the Americas and the euphorbias in Africa represent convergent adaptations to arid environments.

Challenges and Controversies

While adaptive radiation is a well-documented phenomenon, it is not without its challenges and controversies. One key question is the role of ecological opportunity versus intrinsic factors such as genetic variation and developmental constraints in driving diversification.

Furthermore, the definition and identification of adaptive radiation events can be subjective and dependent on the scale of analysis. Some researchers argue that many purported examples of adaptive radiation may instead represent instances of gradual diversification over long periods of time.

Adaptive radiation stands as one of the most remarkable examples of evolutionary innovation and diversification in the history of life on Earth. From the finches of the Galápagos Islands to the marsupials of Australia and the cacti of the Americas, adaptive radiation has shaped the diversity of life in myriad ways.

By exploiting ecological opportunities and evolving key innovations, organisms have diversified and thrived in a wide range of environments, leaving behind a rich tapestry of evolutionary history. In the chapters that follow, we will continue to explore the processes of evolution and adaptation that have shaped the remarkable diversity of life on Earth.

Chapter 11: Coevolution: Interactions Between Species

In the intricate web of life on Earth, species do not exist in isolation but are instead interconnected through a myriad of interactions. Coevolution—the reciprocal evolution of two or more species as a result of their interactions—plays a central role in shaping the diversity and complexity of ecosystems. In this chapter, we delve into the fascinating world of coevolution, exploring the diverse ways in which species influence and shape each other's evolutionary trajectories.

Understanding Coevolution

Coevolution occurs when two or more species exert selective pressures on each other, leading to reciprocal adaptations over time. These interactions can take many forms, including predation, herbivory, mutualism, parasitism, and competition. Coevolutionary processes can drive rapid evolutionary change and result in the diversification and specialization of species.

The Red Queen Hypothesis

The Red Queen hypothesis, named after the character in Lewis Carroll's "Through the Looking-Glass," posits that coevolutionary interactions are akin to a never-ending arms race, with each species constantly evolving to outcompete or evade its adversaries. This metaphor highlights the dynamic and ever-changing nature of coevolutionary relationships, where no species remains static or unchallenged for long.

Predator-Prey Coevolution

One of the most classic examples of coevolution is the arms race between predators and their prey. Predators evolve adaptations for capturing and consuming prey, while prey species evolve defenses to avoid detection and capture. This coevolutionary dynamic can lead to the evolution of elaborate behaviors, morphological traits, and chemical defenses in both predators and prey.

A famous example of predator-prey coevolution is the evolutionary arms race between cheetahs and antelope. Cheetahs have evolved incredible speed and agility to catch their prey, while antelope have evolved swift running abilities and keen senses to evade capture.

Plant-Herbivore Coevolution

Plants and herbivores engage in a perpetual struggle for survival, with each evolving strategies to outwit the other. Plants may evolve physical defenses such as thorns, spines, and tough leaves, or chemical defenses such as toxins and secondary metabolites. In response, herbivores may evolve detoxification mechanisms or specialized feeding behaviors to overcome plant defenses.

The coevolutionary arms race between plants and herbivores has led to the remarkable diversity of plant secondary metabolites, which play a crucial role in mediating interactions between plants and herbivores. For example, the evolution of cyanogenic glycosides in plants has been driven by selection pressure from herbivores, while herbivores have evolved mechanisms to detoxify these compounds and exploit them as food sources.

Mutualistic Coevolution

In mutualistic relationships, two species interact in a mutually beneficial manner, with each providing resources or services that the other requires for survival or reproduction. Mutualistic coevolution often involves the reciprocal evolution of traits that enhance the fitness of both partners.

A classic example of mutualistic coevolution is the relationship between flowering plants and their pollinators. Flowering plants have evolved a diverse array of floral traits, such as color, scent, and morphology, to attract pollinators such as bees, butterflies, and birds. In turn, pollinators have evolved specialized mouthparts, behaviors, and sensory abilities to efficiently extract nectar and pollen from flowers.

Parasite-Host Coevolution

Parasites and their hosts engage in a coevolutionary arms race, with parasites evolving strategies to exploit host resources and hosts evolving defenses to resist parasitic infection. This coevolutionary dynamic can lead to the evolution of elaborate immune systems, behavioral adaptations, and physiological responses in both parasites and hosts.

For example, the arms race between malaria parasites and their human hosts has led to the evolution of genetic variants that confer resistance to malaria infection, such as the sickle cell trait and the Duffy-negative blood group. Similarly, parasites have evolved mechanisms to evade host immune responses and manipulate host behavior to enhance their transmission.

Consequences of Coevolution

Coevolutionary interactions have profound consequences for the ecology, evolution, and biodiversity of ecosystems. These interactions can drive speciation, promote diversification, and shape the structure and dynamics of ecological communities. Furthermore, coevolutionary processes can influence ecosystem stability, resilience, and function, with cascading effects on ecosystem services and human well-being.

Coevolution stands as a testament to the interconnectedness and dynamism of life on Earth, revealing the intricate dance of adaptation and counter-adaptation that shapes the diversity and complexity of ecosystems. From the arms race between predators and prey to the mutualistic partnerships between plants and their pollinators, coevolutionary interactions have left an indelible mark on the evolutionary history of life. In the chapters that follow, we will continue to explore the myriad ways in which species interact and influence each other's evolutionary trajectories, uncovering the hidden connections that bind the web of life together.

Chapter 12: Evolutionary Developmental Biology

Evolutionary developmental biology, or evo-devo, represents a synthesis of developmental biology and evolutionary biology, seeking to understand how changes in developmental processes contribute to the evolution of morphology, anatomy, and behavior. In this chapter, we delve into the fascinating field of evo-devo, exploring the mechanisms of development and their implications for the history of life on Earth.

The Origins of Evo-Devo

The roots of evo-devo can be traced back to the pioneering work of developmental biologists such as Ernst Haeckel, Richard Woltereck, and Conrad Hal Waddington, who sought to understand the processes underlying embryonic development and pattern formation. However, it was not until the latter half of the 20th century that the field began to coalesce as a distinct discipline.

The advent of molecular genetics and advances in imaging technologies have revolutionized our understanding of developmental processes, allowing scientists to unravel the genetic and molecular mechanisms that govern embryonic development. This integration of developmental and evolutionary biology has led to new insights into the processes of evolution and diversification.

Hox Genes and Developmental Patterning

One of the key discoveries in evo-devo is the role of Hox genes in controlling the development of body plans and segmental patterning in animals. Hox genes encode transcription factors that regulate the expression of downstream genes involved in cell fate specification and differentiation.

The spatial and temporal expression of Hox genes along the body axis plays a crucial role in determining the identity of body segments and the formation of specialized structures such as limbs, appendages, and organs. Changes in the expression patterns of Hox genes can lead to alterations in body morphology and contribute to the evolution of novel traits.

Evolution of Developmental Modules

Evo-devo research has revealed that developmental processes are modular, with discrete genetic and molecular pathways controlling different aspects of development. These developmental modules can be co-opted, modified, or repurposed during evolution to generate phenotypic diversity and innovation.

For example, the genetic toolkit for building eyes is highly conserved across diverse animal taxa, indicating that the evolution of eyes has occurred through the modification and redeployment of pre-existing developmental pathways. Similarly, the evolution of wings in insects and birds has involved the co-option of limb development genes to specify wing identity and morphology.

Evo-Devo and Evolutionary Transitions

Evo-devo research has shed light on the evolutionary transitions that have shaped the history of life on Earth, from the evolution of multicellularity and body plans to the origin of novel traits and adaptations. By studying the genetic and developmental mechanisms underlying these transitions, scientists can gain insights into the processes driving evolutionary change.

One classic example is the evolution of the vertebrate limb, which has undergone multiple transformations over millions of years, giving rise to diverse forms such as fins, wings, and limbs. Comparative studies of limb development in different vertebrate species have revealed conserved genetic pathways controlling limb patterning and growth, as well as species-specific modifications that contribute to morphological diversity.

Evo-Devo and Evolutionary Constraints

Evo-devo research has also shed light on the role of developmental constraints in shaping the trajectory of evolution. Developmental constraints arise from the limitations imposed by the structure, organization, and regulatory networks of developmental systems, which can restrict the range of phenotypic variation that is accessible to natural selection.

For example, the bauplan, or body plan, of animals imposes constraints on the types of structures and morphologies that can evolve within a particular lineage. Similarly, the genetic and developmental interactions that underlie complex traits such as organ systems and physiological processes may constrain the evolution of certain phenotypes.

Applications of Evo-Devo

Evo-devo research has practical applications in fields such as medicine, agriculture, and conservation, where understanding the genetic and developmental basis of traits is crucial for addressing human health, food security, and biodiversity conservation challenges. By elucidating the mechanisms of development and evolution, evo-devo provides a foundation for studying the origins of genetic diseases, the breeding of crops and livestock, and the conservation of endangered species.

Evolutionary developmental biology offers a powerful framework for understanding the interplay between development and evolution, shedding light on the processes that have shaped the diversity and complexity of life on Earth. From the genetic regulation of embryonic development to the evolution of novel traits and adaptations, evo-devo provides a window into the deep connections between form and function across the tree of life. In the chapters that follow, we will continue to explore the myriad ways in which genetic and developmental processes have shaped the evolutionary history of organisms and ecosystems.

Chapter 13: Speciation: Origin of New Species

Speciation, the process by which new species arise, lies at the heart of evolutionary biology. From the divergence of populations to the formation of distinct lineages, speciation drives the diversification and adaptation of life on Earth. In this chapter, we explore the mechanisms of speciation and the factors that contribute to the origin of new species.

The Concept of Species

The concept of species is central to the study of speciation, yet defining what constitutes a species has long been a subject of debate among biologists. Traditionally, species have been defined based on morphological, ecological, or reproductive criteria, with the biological species concept, proposed by Ernst Mayr, emphasizing reproductive isolation as the defining criterion.

According to the biological species concept, species are groups of organisms that are reproductively isolated from one another, meaning that they do not interbreed and produce viable offspring under natural conditions. While this definition is widely accepted, it does not apply to all organisms, particularly those that reproduce asexually or have complex mating systems.

Modes of Speciation

Speciation can occur through a variety of mechanisms, with two main modes: allopatric speciation and sympatric speciation. Allopatric speciation occurs when populations become geographically isolated from one another, leading to reproductive isolation and the divergence of genetic traits over time.

One classic example of allopatric speciation is the formation of new species of finches in the Galápagos Islands, where different island populations became isolated from each other and evolved unique beak shapes and feeding habits. Sympatric speciation, on the other hand, occurs when populations diverge into separate species within the same geographic area, often as a result of ecological or behavioral factors.

Reinforcement and Reproductive Isolation

Reproductive isolation plays a crucial role in the speciation process, preventing gene flow between populations and promoting divergence and reproductive compatibility. Reinforcement is a process whereby natural selection favors traits that reduce the likelihood of hybridization between diverging populations, reinforcing reproductive isolation and promoting speciation.

Reinforcement often occurs in secondary contact zones, where previously isolated populations come into contact and hybridize. Natural selection acts to strengthen prezygotic barriers to mating and reproduction, such as differences in mating signals, courtship behaviors, or habitat preferences, reducing the likelihood of hybridization and promoting the evolution of distinct species.

Hybridization and Speciation

While reproductive isolation typically acts as a barrier to gene flow between species, hybridization can sometimes occur between divergent lineages, leading to the formation of hybrid zones and the exchange of genetic material. In some cases, hybridization may facilitate speciation by introducing novel genetic variation and promoting adaptive divergence.

One classic example of hybrid speciation is the formation of new species of sunflowers in North America, where hybridization between different species led to the origin of fertile hybrid lineages with unique combinations of traits. Hybrid speciation challenges traditional views of speciation as a gradual, stepwise process and highlights the role of hybridization in generating biodiversity.

Ecological Speciation

Ecological speciation occurs when diverging populations adapt to different ecological niches, leading to reproductive isolation and the formation of new species. Ecological factors such as habitat preference, resource use, and ecological interactions can drive divergent selection and promote reproductive isolation between populations.

One classic example of ecological speciation is the formation of new species of stickleback fish in freshwater lakes, where divergent selection pressures such as predation, competition, and habitat complexity have led to the evolution of distinct morphological and behavioral traits. Ecological speciation highlights the importance of environmental factors in driving evolutionary divergence and promoting the origin of new species.

Hybrid Zones and Speciation Continua

Hybrid zones provide natural laboratories for studying the process of speciation and the dynamics of gene flow between divergent populations. In hybrid zones, hybridization between closely related species can lead to the formation of hybrid swarms, where genetic material is exchanged and introgressed between populations.

Hybrid zones can also serve as natural barriers to gene flow, promoting reproductive isolation and the formation of new species. The structure and dynamics of hybrid zones vary depending on factors such as habitat heterogeneity, dispersal patterns, and the strength of selection against hybrids, providing insights into the mechanisms driving speciation.

Speciation represents a fundamental process in the history of life on Earth, driving the diversification and adaptation of organisms across diverse habitats and ecosystems. From the isolation of populations to the evolution of reproductive barriers, speciation occurs through a variety of mechanisms and pathways, shaping the tree of life in myriad ways.

By studying the mechanisms and patterns of speciation, scientists can gain insights into the processes driving evolutionary change and the origins of biodiversity. In the chapters that follow, we will continue to explore the dynamic interplay between genetic, ecological, and evolutionary processes, uncovering the hidden mechanisms that underlie the origin and diversification of life on Earth.

Chapter 14: Human Evolution: From Australopithecus to Homo sapiens

The story of human evolution is one of the most captivating narratives in the history of life on Earth. From our humble beginnings as hominins in Africa to the emergence of Homo sapiens as the dominant species on the planet, the journey of human evolution is marked by innovation, adaptation, and resilience. In this chapter, we embark on a journey through time, tracing the footsteps of our ancestors and exploring the remarkable transformations that have shaped the lineage of Homo sapiens.

The Origins of Hominins

The story of human evolution begins in Africa, where our earliest ancestors, known as hominins, first appeared more than six million years ago. The earliest hominins were small-bodied, ape-like creatures adapted to life in the dense forests of Africa. Over time, these early hominins began to adapt to new environments and ecological niches, leading to the diversification of the hominin lineage.

One of the earliest hominins known to science is Sahelanthropus tchadensis, whose fossilized remains were discovered in Chad and dated to approximately seven million years ago. Sahelanthropus exhibits a unique combination of ape-like and human-like features, making it a key candidate for the last common ancestor of humans and chimpanzees.

Australopithecus: The Southern Ape

The genus Australopithecus represents a critical stage in human evolution, marked by significant changes in anatomy, behavior, and ecology. Australopithecus species were bipedal hominins that lived in Africa between approximately four and two million years ago. They were characterized by a combination of ape-like and human-like features, including small brains, large jaws, and adaptations for upright walking.

One of the most famous Australopithecus species is Australopithecus afarensis, whose fossilized remains were discovered in Ethiopia and Tanzania. Australopithecus afarensis, famously represented by the fossil "Lucy," provides valuable insights into the anatomy and behavior of early hominins, including their locomotion, diet, and social structure.

The Genus Homo: Early Toolmakers

The emergence of the genus Homo marked a significant milestone in human evolution, characterized by the evolution of larger brains, reduced jaws, and increased reliance on tool use. The earliest members of the genus Homo, such as Homo habilis and Homo rudolfensis, appeared in Africa around two million years ago and were distinguished by their ability to manufacture and use stone tools.

Homo habilis, known as the "handy man," was the first hominin species to be associated with stone tools, suggesting an increased reliance on technology for accessing and processing food resources. Homo habilis coexisted with Australopithecus species for a time, but eventually gave rise to more advanced members of the genus Homo.

Homo erectus: The First Global Explorers

Homo erectus represents one of the most successful and widespread hominin species in the history of human evolution. Appearing in Africa around 1.8 million years ago, Homo erectus was characterized by its tall stature, long legs, and large brains, as well as its ability to adapt to a wide range of environments.

Homo erectus was the first hominin species to migrate out of Africa, spreading across Eurasia and colonizing diverse habitats from savannas to woodlands. Their ability to control fire, make more sophisticated tools, and adapt to changing climates allowed Homo erectus to thrive in a variety of environments and eventually give rise to other hominin lineages.

The Neanderthals: Our Closest Relatives

The Neanderthals, or Homo neanderthalensis, represent one of the closest relatives of modern humans, sharing a common ancestor with Homo sapiens around 600,000 years ago. Neanderthals were adapted to life in Ice Age Europe, with robust bodies, large brains, and specialized adaptations for hunting and survival in cold climates.

Neanderthals were skilled toolmakers and hunters, using tools such as spears, knives, and scrapers to hunt game and process food. They also exhibited complex social behaviors, including care for the sick and injured, burial of the dead, and symbolic expression through art and ornamentation.

The Emergence of Homo sapiens

The origin of our own species, Homo sapiens, is a topic of ongoing debate and research among paleoanthropologists. Genetic and fossil evidence suggest that Homo sapiens evolved in Africa around 300,000 to 200,000 years ago, possibly from a population of archaic Homo sapiens or Homo heidelbergensis.

The earliest known fossils of anatomically modern humans come from sites in Africa, including Jebel Irhoud in Morocco and Omo Kibish in Ethiopia. These fossils exhibit the characteristic features of Homo sapiens, including a globular braincase, reduced brow ridges, and a prominent chin, indicating the emergence of our own species.

Out of Africa: The Spread of Homo sapiens

Around 70,000 to 60,000 years ago, Homo sapiens began to migrate out of Africa and spread across the globe, eventually colonizing every continent except Antarctica. This dispersal, known as the Out of Africa migration, represents one of the most significant events in human history, shaping the genetic diversity and cultural evolution of modern humans.

The migration of Homo sapiens was facilitated by a combination of technological innovation, environmental change, and social adaptation. Modern humans used boats and coastal routes to reach distant islands and continents, eventually reaching places as far-flung as Australia, the Americas, and remote Pacific islands.

Interactions with Other Hominins

As Homo sapiens spread across the globe, they encountered and interacted with other hominin species such as the Neanderthals and Denisovans. Genetic evidence suggests that modern humans interbred with these archaic hominins, leading to the incorporation of Neanderthal and Denisovan DNA into the modern human gene pool.

These interbreeding events highlight the complex and dynamic nature of human evolution, as well as the interconnectedness of different hominin lineages. They also raise questions about the nature of human diversity and the role of gene flow in shaping the evolutionary history of our species.

The story of human evolution is a testament to the power of adaptation, innovation, and resilience in the face of environmental challenges. From our earliest ancestors in Africa to the spread of Homo sapiens across the globe, the journey of human evolution is marked by diversity, complexity, and interconnectedness.

By studying the fossil record, genetic evidence, and archaeological remains, scientists continue to unravel the mysteries of human evolution, shedding light on our origins, our place in the natural world, and the forces that have shaped our species over millions of years.

Chapter 15: Evolutionary Medicine: Applying Evolutionary Principles to Health

Evolutionary medicine is a field that applies principles of evolutionary biology to understand the origins and mechanisms of human diseases and to inform medical practice and public health interventions. By examining the evolutionary history of humans and the pathogens that affect them, evolutionary medicine offers insights into the underlying causes of disease and informs strategies for prevention, treatment, and health promotion. In this chapter, we explore the key concepts and applications of evolutionary medicine, highlighting its role in improving human health and well-being.

The Evolutionary Basis of Disease

Many of the diseases that afflict humans have cvolutionary origins, shaped by the interactions between hosts, pathogens, and the environment over millions of years. Infectious diseases, in particular, have played a significant role in human evolution, driving the selection of genetic variants that confer resistance or susceptibility to infection.

For example, the genetic diversity of the human leukocyte antigen (HLA) system, which plays a crucial role in immune response, is thought to have evolved in response to the selective pressure exerted by pathogens such as viruses, bacteria, and parasites. Similarly, traits such as sickle cell hemoglobin and glucose-6-phosphate dehydrogenase deficiency have evolved as adaptations to malaria and other infectious diseases.

The Hygiene Hypothesis

The hygiene hypothesis posits that the modern increase in allergic and autoimmune diseases may be partly attributable to reduced exposure to infectious agents during childhood, leading to dysregulation of the immune system. According to this hypothesis, exposure to a diverse array of microbes during early development helps to train and calibrate the immune system, reducing the risk of allergic and autoimmune reactions later in life.

Support for the hygiene hypothesis comes from studies showing higher rates of allergic and autoimmune diseases in developed countries compared to developing countries, as well as from experimental studies in animals demonstrating the protective effects of early microbial exposure. These findings underscore the importance of maintaining a balanced microbiome and immune system for optimal health.

Evolutionary Mismatch and Chronic Disease

Evolutionary mismatch occurs when modern environments and lifestyles diverge from the conditions under which human ancestors evolved, leading to mismatches between our biological adaptations and our current environments. This mismatch can contribute to the development of chronic diseases such as obesity, diabetes, cardiovascular disease, and mental health disorders.

For example, the mismatch between a diet high in processed foods and sedentary lifestyles and our evolutionary adaptations for energy storage and expenditure may contribute to the epidemic of obesity and metabolic syndrome. Similarly, the mismatch between artificial light exposure and natural circadian rhythms may disrupt sleep-wake cycles and contribute to sleep disorders and mood disturbances.

Evolutionary Trade-offs and Disease

Evolutionary trade-offs occur when adaptations to one aspect of the environment incur costs or trade-offs in other aspects of health and fitness. These trade-offs can manifest at the genetic, physiological, or behavioral level and may influence susceptibility to disease and the effectiveness of medical interventions.

For example, the thrifty genotype hypothesis proposes that genetic variants that confer resistance to starvation and energy deprivation may increase the risk of obesity and metabolic syndrome in environments of food abundance. Similarly, adaptations to high-altitude environments, such as increased red blood cell production, may increase the risk of cardiovascular and respiratory diseases at lower altitudes.

Evolutionary Approaches to Disease Prevention and Treatment

Evolutionary principles can inform strategies for disease prevention, treatment, and health promotion by targeting the underlying causes of disease and leveraging natural selection and adaptation. For example, evolutionary insights into the origins and spread of infectious diseases can inform vaccination strategies, antimicrobial stewardship programs, and surveillance efforts to monitor emerging pathogens.

Similarly, understanding the evolutionary mechanisms of drug resistance can guide the development of new antimicrobial agents and combination therapies to combat resistant strains of bacteria, viruses, and parasites. Evolutionary medicine also emphasizes the importance of personalized and precision medicine approaches that take into account individual genetic variation, environmental exposures, and lifestyle factors.

Evolutionary medicine offers a powerful framework for understanding the origins and mechanisms of human diseases and for developing innovative strategies for disease prevention, treatment, and health promotion. By examining the evolutionary history of humans and the pathogens that affect them, evolutionary medicine provides insights into the complex interactions between hosts, pathogens, and the environment.

By applying evolutionary principles to health, we can improve our understanding of disease processes, optimize medical interventions, and enhance the health and well-being of individuals and populations. In the chapters that follow, we will continue to explore the dynamic interplay between evolution, biology, and medicine, uncovering the hidden connections that shape the health and diversity of life on Earth.

Chapter 16: Conservation Biology: Evolutionary Approaches to Biodiversity Conservation

Conservation biology is a multidisciplinary field that aims to understand and preserve the diversity of life on Earth. By integrating principles of ecology, genetics, and evolutionary biology, conservation biologists develop strategies for protecting species, habitats, and ecosystems from threats such as habitat loss, climate change, pollution, and invasive species. In this chapter, we explore the evolutionary approaches and principles that underpin biodiversity conservation, highlighting their importance in safeguarding the natural world for future generations.

Understanding Biodiversity

Biodiversity encompasses the variety of life on Earth, including species diversity, genetic diversity, and ecosystem diversity. Conservation biologists recognize the importance of preserving biodiversity not only for its intrinsic value but also for the ecosystem services it provides, such as pollination, water purification, and climate regulation.

One of the pioneering figures in conservation biology is Edward O. Wilson, whose work on biodiversity and ecosystem function has highlighted the critical role of species interactions in maintaining ecosystem stability and resilience. Wilson's concept of "biophilia" emphasizes the innate human connection to nature and the importance of preserving biodiversity for human well-being.

Evolutionary Perspectives on Conservation

Evolutionary biology provides valuable insights into the processes that shape patterns of biodiversity and the factors that influence species persistence and extinction. By understanding the evolutionary history of species, conservation biologists can identify key evolutionary processes such as adaptation, speciation, and extinction that influence conservation outcomes.

One of the central tenets of evolutionary biology is the concept of phylogenetic diversity, which considers the evolutionary relationships among species when prioritizing conservation efforts. Phylogenetic diversity reflects the unique evolutionary history of different lineages and can help identify conservation priorities that maximize evolutionary distinctiveness and representativeness.

Conservation Genetics

Conservation genetics is a subfield of conservation biology that uses genetic data to inform conservation strategies and management decisions. By studying patterns of genetic variation within and among populations, conservation geneticists can assess levels of genetic diversity, detect inbreeding and genetic bottlenecks, and identify populations at risk of extinction.

The pioneering work of conservation geneticists such as Richard Frankham and Michael Soulé has highlighted the importance of genetic diversity for population viability and long-term survival. Genetic techniques such as DNA sequencing, microsatellite analysis, and population genomics have revolutionized our ability to study and conserve endangered species.

Evolutionary Rescue and Adaptation

Evolutionary rescue occurs when populations are able to adapt to rapid environmental change through natural selection and genetic evolution. By studying the evolutionary potential of populations, conservation biologists can identify strategies for promoting adaptive responses to environmental stressors such as climate change, habitat fragmentation, and pollution.

One example of evolutionary rescue is the adaptation of certain fish populations to pollution in industrialized rivers, where genetic variants conferring resistance to toxins have become more common over time. Understanding the genetic basis of adaptation can inform conservation efforts aimed at restoring and maintaining the evolutionary potential of populations.

Assisted Evolution and Genetic Rescue

Assisted evolution involves the deliberate manipulation of evolutionary processes to enhance the resilience and adaptive capacity of populations facing environmental threats. Techniques such as selective breeding, translocation of individuals, and genetic augmentation can be used to introduce genetic variation or adaptive traits into threatened populations.

Genetic rescue is a specific form of assisted evolution that involves introducing new genetic material into small or inbred populations to increase genetic diversity and improve population fitness. Genetic rescue efforts have been successful in restoring the viability of endangered species such as the Florida panther and the black-footed ferret.

Conservation Paleobiology

Conservation paleobiology is an emerging field that uses the fossil record to inform conservation strategies and management decisions. By studying the responses of species and ecosystems to past environmental changes, conservation paleobiologists can gain insights into the long-term dynamics of biodiversity and the potential effects of future climate change.

The study of "paleoecological analogs" provides a valuable tool for predicting how species and ecosystems may respond to ongoing environmental changes. By comparing past and present environments, conservation paleobiologists can identify areas of conservation priority and develop strategies for mitigating the impacts of climate change on biodiversity.

Integrating Evolutionary Approaches into Conservation Practice

Integrating evolutionary approaches into conservation practice requires collaboration among scientists, policymakers, and stakeholders to develop evidence-based strategies for protecting biodiversity. By combining ecological, genetic, and evolutionary data, conservation practitioners can prioritize conservation actions that maximize the adaptive capacity and resilience of species and ecosystems.

One example of successful integration is the "Evolutionarily Significant Unit" (ESU) framework, which recognizes the evolutionary distinctiveness of populations when designating conservation units. ESUs consider both genetic and ecological criteria to identify populations that are evolutionarily significant and warrant special conservation attention.

Evolutionary approaches play a critical role in shaping biodiversity conservation strategies and informing management decisions in a rapidly changing world. By understanding the evolutionary processes that govern patterns of biodiversity, conservation biologists can develop more effective and sustainable solutions for protecting the natural world.

As we continue to face unprecedented challenges such as habitat loss, climate change, and species extinction, the insights provided by evolutionary biology will be essential for guiding conservation efforts and preserving the rich tapestry of life on Earth for future generations.

Chapter 17: Evolutionary Ecology: Interactions Between Organisms and Their Environment

Evolutionary ecology is a field that explores the dynamic interplay between organisms and their environment, focusing on how ecological interactions shape evolutionary processes and patterns of biodiversity. By studying the evolutionary responses of organisms to environmental pressures such as competition, predation, and resource availability, evolutionary ecologists gain insights into the mechanisms driving adaptation, speciation, and community dynamics. In this chapter, we delve into the fascinating world of evolutionary ecology, exploring the complex relationships between organisms and their surroundings and their implications for the diversity and stability of ecosystems.

The Evolutionary Basis of Ecology

The study of ecology examines the interactions between organisms and their environment, including factors such as climate, habitat, and other species. Evolutionary ecology builds upon this foundation by considering how these interactions have shaped the evolution of traits, behaviors, and life histories across diverse taxa.

One of the pioneers of evolutionary ecology is G. Evelyn Hutchinson, whose research on niche theory and ecological interactions laid the groundwork for understanding the dynamics of ecological communities. Hutchinson's concept of the ecological niche, defined as the set of environmental conditions in which a species can persist and reproduce, remains central to the study of evolutionary ecology.

Adaptation to the Environment

Adaptation is the process by which organisms evolve traits that enhance their survival and reproduction in a particular environment.

Evolutionary ecologists study how natural selection acts on heritable variation within populations to produce adaptive phenotypes that are better suited to their ecological niche.

Classic examples of adaptation include the evolution of camouflage in prey species to avoid predation, the development of specialized feeding structures in pollinators and herbivores, and the acquisition of physiological adaptations for coping with extreme environments such as deserts, mountains, and aquatic habitats.

Coevolutionary Interactions

Coevolution occurs when two or more species interact closely with one another and exert reciprocal selective pressures on each other's traits. Coevolutionary interactions can lead to the evolution of specialized adaptations, such as mimicry, mutualism, and antagonistic defenses, as species adapt to one another's presence and behaviors.

One iconic example of coevolution is the arms race between predators and prey, where prey species evolve defensive mechanisms such as chemical defenses, camouflage, and warning coloration to evade predation, while predators evolve counter-adaptations for capturing and consuming prey.

Species Interactions and Community Dynamics

Species interactions such as competition, predation, and mutualism play a crucial role in shaping the structure and dynamics of ecological communities. By studying the patterns and outcomes of species interactions, evolutionary ecologists can gain insights into the factors that influence species coexistence, diversity, and distribution.

For example, the competitive exclusion principle, proposed by Gause, states that two species competing for the same limiting resource cannot coexist indefinitely, leading to the exclusion of one species by the other. However, in practice, many species are able to coexist through mechanisms such as niche differentiation, resource partitioning, and spatial or temporal segregation.

Evolutionary Trade-offs and Constraints

Evolutionary trade-offs occur when selection for one trait comes at the expense of another, leading to compromises in organismal performance or fitness. These trade-offs can arise from genetic, physiological, or ecological constraints and may influence the evolution of traits such as reproductive effort, lifespan, and dispersal ability.

For example, the evolution of larger body size may confer advantages in competitive interactions or predator avoidance but may also increase energetic costs and vulnerability to environmental stressors. Similarly, the allocation of resources to reproduction may trade off against investment in somatic maintenance and longevity, shaping life history strategies and population dynamics.

Ecological Resilience and Stability

Ecological resilience refers to the ability of ecosystems to absorb and recover from disturbances while maintaining their structure, function, and integrity. Evolutionary processes such as genetic diversity, phenotypic plasticity, and adaptive capacity can enhance the resilience of populations and communities to environmental change.

By studying the evolutionary responses of organisms to perturbations such as habitat loss, climate change, and invasive species, evolutionary ecologists can assess the adaptive potential of species and ecosystems and develop strategies for enhancing their resilience and stability over time.

Conservation Implications

Evolutionary ecology has important implications for biodiversity conservation and ecosystem management, particularly in the face of global environmental change. By understanding the evolutionary processes that govern patterns of biodiversity and ecosystem function, conservation practitioners can develop more effective strategies for protecting and restoring natural habitats and populations.

For example, incorporating evolutionary principles into conservation planning can help identify priority areas for conservation, predict the responses of species to environmental change, and design interventions that promote adaptation and genetic diversity. By considering the adaptive potential of species and ecosystems, conservation efforts can become more proactive and anticipatory, ensuring the long-term viability of Earth's biodiversity.

Evolutionary ecology provides a powerful framework for understanding the dynamic interactions between organisms and their environment and for predicting the responses of species and ecosystems to environmental change. By studying the evolutionary processes that underpin ecological dynamics, evolutionary ecologists can uncover the hidden mechanisms that shape the diversity, resilience, and stability of life on Earth.

As we continue to grapple with pressing environmental challenges such as habitat loss, climate change, and species extinction, the insights provided by evolutionary ecology will be essential for guiding conservation efforts and preserving the natural world for future generations.

Conclusion: Exploring the Tapestry of Life

As we conclude our journey through the captivating landscape of evolutionary biology, we are reminded of the intricate tapestry of life that spans billions of years of Earth's history. From the origins of life in ancient oceans to the emergence of complex ecosystems and diverse forms of life, the story of evolution is a testament to the power of adaptation, innovation, and resilience in the face of environmental change.

Throughout this book, we have traced the history of life on Earth, exploring the dynamic processes and patterns that have shaped the diversity and complexity of living organisms. We have delved into the mechanisms of evolution, from natural selection and genetic drift to speciation and adaptation, uncovering the hidden mechanisms that drive evolutionary change.

We have marveled at the wonders of the fossil record, where the preserved remains of ancient organisms offer glimpses into past worlds and lost ecosystems. From the Burgess Shale of Canada to the Ediacaran biota of Australia, fossil sites around the globe provide windows into the origins and evolution of life on Earth.

We have explored the tree of life, tracing the evolutionary relationships among organisms and uncovering the hidden connections that bind us to our distant relatives. From bacteria to fungi, plants to animals, the diversity of life on Earth is a testament to the power of evolution to generate and sustain biological complexity.

We have delved into the realm of molecular biology, where advances in DNA sequencing and genomics have revolutionized our understanding of evolutionary processes. From the sequencing of the human genome to the study of ancient DNA from Neanderthals and Denisovans, molecular techniques continue to shed light on the genetic basis of evolution and the origins of biodiversity.

We have examined the intersections between evolution and other fields of science, from ecology and conservation biology to medicine and agriculture. By applying evolutionary principles to real-world problems, scientists and practitioners are developing innovative solutions for addressing global challenges such as climate change, habitat loss, and emerging infectious diseases.

As we reflect on the insights and discoveries of evolutionary biology, we are reminded of the importance of curiosity, inquiry, and exploration in advancing our understanding of the natural world. From the shores of the Galápagos Islands to the halls of academic institutions and research laboratories, scientists and scholars continue to push the boundaries of knowledge and expand our horizons of understanding.

But our journey does not end here. As we look to the future, we are faced with new questions and challenges that demand our attention and ingenuity. How will ecosystems respond to rapid environmental change? What role will humans play in shaping the course of evolution? How can we harness the power of evolution to address pressing societal needs and ensure the sustainability of life on Earth?

These are questions that will guide the next chapter of our journey, as we continue to explore the wonders of evolutionary biology and trace the history of life on Earth. With each new discovery and each new insight, we come closer to unraveling the mysteries of existence and unlocking the secrets of our shared heritage.

As we bid farewell to the pages of this book, let us carry forward the spirit of inquiry and discovery that has fueled our exploration of the natural world. Let us continue to marvel at the beauty and complexity of life on Earth and to cherish the remarkable journey that has brought us to this moment in time.

For in the end, it is our curiosity, our passion, and our dedication to understanding the world around us that will drive us forward on our quest to unlock the secrets of evolution and to trace the history of life on Earth.

So let us embark on this journey together, with open minds and open hearts, as we explore the wonders of evolutionary biology and celebrate the rich tapestry of life that surrounds us.

Thank you for joining us on this extraordinary adventure. Until we meet again, may your curiosity never wane, and may the wonders of evolution continue to inspire and awe us all.

Acknowledgements

Completing this book, "Evolutionary Biology: Tracing the History of Life on Earth," has been a journey filled with discovery, reflection, and collaboration. We are immensely grateful to all those who have contributed their time, expertise, and support to make this project possible.

First and foremost, we extend our heartfelt gratitude to the pioneers and visionaries of evolutionary biology whose groundbreaking research has paved the way for our understanding of the natural world. From Charles Darwin and Alfred Russel Wallace to G. Evelyn Hutchinson and E. O. Wilson, their insights and discoveries continue to inspire generations of scientists and scholars.

We are deeply thankful to the countless researchers, educators, and practitioners whose work has expanded our knowledge of evolutionary processes and their implications for life on Earth. From field expeditions to laboratory experiments, their dedication and passion for discovery have enriched our understanding of the complexities of evolution.

We would like to express our appreciation to the institutions and organizations that have supported our research and provided resources for the completion of this book. From universities and research institutes to funding agencies and libraries, their contributions have been invaluable in advancing the frontiers of evolutionary biology.

We are indebted to our colleagues and collaborators who have shared their insights, expertise, and feedback throughout the writing process. Their thoughtful comments and constructive criticism have helped shape the content and clarity of this book, ensuring its accuracy and relevance to the field of evolutionary biology.

We extend our gratitude to the editorial and production teams who have worked tirelessly behind the scenes to bring this book to fruition. Their dedication, professionalism, and attention to detail have been instrumental in turning our vision into reality and ensuring the highest quality standards for our readers.

Finally, we would like to thank our families, friends, and loved ones for their unwavering support, encouragement, and understanding throughout the writing process. Their patience, encouragement, and love have been our constant source of inspiration and motivation, reminding us of the importance of curiosity, perseverance, and resilience in the pursuit of knowledge.

To all those who have contributed to this book in ways large and small, we extend our deepest gratitude. May the pages of "Evolutionary Biology: Tracing the History of Life on Earth" inspire curiosity, ignite passion, and spark new discoveries in the minds of readers around the world.

With heartfelt thanks,

Charlie O. Williams

www.ingramcontent.com/pod-product-compliance
Lightning Source LLC
Chambersburg PA
CBHW050109230526
45470CB00004B/1745